U.S. Department of Transportation
National Highway Traffic Safety Administration

DOT HS 811 499　　　　　　　　　　　　　　　　　　July 2011

System Capability Assessment of Cooperative Intersection Collision Avoidance System for Violations (CICAS-V)

DISCLAIMER

This publication is distributed by the U.S. Department of Transportation, National Highway Traffic Safety Administration, in the interest of information exchange. The opinions, findings, and conclusions expressed in this publication are those of the authors and not necessarily those of the Department of Transportation or the National Highway Traffic Safety Administration. The United States Government assumes no liability for its contents or use thereof. If trade names, manufacturers' names, or specific products are mentioned, it is because they are considered essential to the object of the publication and should not be construed as an endorsement. The United States Government does not endorse products or manufacturers.

REPORT DOCUMENTATION PAGE		*Form Approved* *OMB No. 0704-0188*

Public reporting burden for this collection of information is estimated to average 1 hour per response, including the time for reviewing instructions, searching existing data sources, gathering and maintaining the data needed, and completing and reviewing the collection of information. Send comments regarding this burden estimate or any other aspect of this collection of information, including suggestions for reducing this burden, to Washington Headquarters Services, Directorate for Information Operations and Reports, 1215 Jefferson Davis Highway, Suite 1204, Arlington, VA 22202-4302, and to the Office of Management and Budget, Paperwork Reduction Project (0704-0188), Washington, DC 20503.

1. AGENCY USE ONLY (Leave blank)	2. REPORT DATE July 2011	3. REPORT TYPE AND DATES COVERED October 2008 – September 2010
4. TITLE AND SUBTITLE System Capability Assessment of Cooperative Intersection Collision Avoidance System for Violations (CICAS-V)		5. FUNDING NUMBERS Inter-Agency Agreement HS-51A1 DTNH22-08-V-00017
6. AUTHOR(S) John Brewer, Jonathan Koopmann, and Wassim G. Najm		
7. PERFORMING ORGANIZATION NAME(S) AND ADDRESS(ES) U.S. Department of Transportation Research and Innovative Technology Administration John A. Volpe National Transportation Systems Center Cambridge, MA 02142		8. PERFORMING ORGANIZATION REPORT NUMBER DOT-VNTSC-NHTSA-11-08
9. SPONSORING/MONITORING AGENCY NAME(S) AND ADDRESS(ES) John Harding U.S. Department of Transportation National Highway Traffic Safety Administration Washington, DC 20590		10. SPONSORING/MONITORING AGENCY REPORT NUMBER DOT HS 811 499
11. SUPPLEMENTARY NOTES		
12a. DISTRIBUTION/AVAILABILITY STATEMENT Document is available to the public from the National Technical Information Service www.ntis.gov		12b. DISTRIBUTION CODE

13. ABSTRACT (Maximum 200 words)

This report describes the system capability assessment for the Cooperative Intersection Collision Avoidance System for Violations (CICAS-V) based on data collected from objective tests and a pilot test. The CICAS-V is a vehicle-to-infrastructure system that provides visual, audio, and haptic (brake pulse) warnings when a vehicle is in danger of violating a traffic signal or stop sign at an intersection. A series of objective tests were conducted at the Virginia Tech Transportation Institute. Each test was defined by an initial geometry, a set of validity constraints (such as the sufficient GPS accuracy), and a set of pass/fail criteria. The test series investigated the ability to appropriately warn or not warn at various speeds, in appropriately discerned lanes, under dynamic lane changes, under changing signal conditions, and in the presence of multiple equipped intersections. A pilot test was run with nearly 100 naïve drivers on a two-hour prescribed course in the Blacksburg, VA area. During the pilot test, drivers were appropriately warned when in danger of violating an obscured stop sign and when intentionally distracted. An algorithm was found to produce occasional nuisance warnings and remedied. An erroneous lane location in a geographical intersection description and a de-synchronized set of roadside equipment also produced some nuisance warnings and were also remedied.

14. SUBJECT TERMS Cooperative Intersection Collision Avoidance System for Violations, CICAS-V, objective tests, pilot test, signaled intersection, stop sign, road side equipment, onboard equipment, Vehicle-to-infrastructure, V2I, GPS, geographical intersection description, GID, lane detection, lane change, Signal Phase and Timing, SPaT, Crash Avoidance Metrics Partnership, CAMP, Dedicated Short Range Communications, DSRC			15. NUMBER OF PAGES 46
			16. PRICE CODE
17. SECURITY CLASSIFICATION OF REPORT Unclassified	18. SECURITY CLASSIFICATION OF THIS PAGE Unclassified	19. SECURITY CLASSIFICATION OF ABSTRACT Unclassified	20. LIMITATION OF ABSTRACT

NSN 7540-01-280-5500

Standard Form 298 (Rev. 2-89)
Prescribed by ANSI Std. 239-18
298-102

METRIC/ENGLISH CONVERSION FACTORS

ENGLISH TO METRIC

LENGTH (APPROXIMATE)
- 1 inch (in) = 2.5 centimeters (cm)
- 1 foot (ft) = 30 centimeters (cm)
- 1 yard (yd) = 0.9 meter (m)
- 1 mile (mi) = 1.6 kilometers (km)

AREA (APPROXIMATE)
- 1 square inch (sq in, in^2) = 6.5 square centimeters (cm^2)
- 1 square foot (sq ft, ft^2) = 0.09 square meter (m^2)
- 1 square yard (sq yd, yd^2) = 0.8 square meter (m^2)
- 1 square mile (sq mi, mi^2) = 2.6 square kilometers (km^2)
- 1 acre = 0.4 hectare (he) = 4,000 square meters (m^2)

MASS - WEIGHT (APPROXIMATE)
- 1 ounce (oz) = 28 grams (gm)
- 1 pound (lb) = 0.45 kilogram (kg)
- 1 short ton = 2,000 pounds (lb) = 0.9 tonne (t)

VOLUME (APPROXIMATE)
- 1 teaspoon (tsp) = 5 milliliters (ml)
- 1 tablespoon (tbsp) = 15 milliliters (ml)
- 1 fluid ounce (fl oz) = 30 milliliters (ml)
- 1 cup (c) = 0.24 liter (l)
- 1 pint (pt) = 0.47 liter (l)
- 1 quart (qt) = 0.96 liter (l)
- 1 gallon (gal) = 3.8 liters (l)
- 1 cubic foot (cu ft, ft^3) = 0.03 cubic meter (m^3)
- 1 cubic yard (cu yd, yd^3) = 0.76 cubic meter (m^3)

TEMPERATURE (EXACT)
$[(x-32)(5/9)]\,°F = y\,°C$

METRIC TO ENGLISH

LENGTH (APPROXIMATE)
- 1 millimeter (mm) = 0.04 inch (in)
- 1 centimeter (cm) = 0.4 inch (in)
- 1 meter (m) = 3.3 feet (ft)
- 1 meter (m) = 1.1 yards (yd)
- 1 kilometer (km) = 0.6 mile (mi)

AREA (APPROXIMATE)
- 1 square centimeter (cm^2) = 0.16 square inch (sq in, in^2)
- 1 square meter (m^2) = 1.2 square yards (sq yd, yd^2)
- 1 square kilometer (km^2) = 0.4 square mile (sq mi, mi^2)
- 10,000 square meters (m^2) = 1 hectare (ha) = 2.5 acres

MASS - WEIGHT (APPROXIMATE)
- 1 gram (gm) = 0.036 ounce (oz)
- 1 kilogram (kg) = 2.2 pounds (lb)
- 1 tonne (t) = 1,000 kilograms (kg) = 1.1 short tons

VOLUME (APPROXIMATE)
- 1 milliliter (ml) = 0.03 fluid ounce (fl oz)
- 1 liter (l) = 2.1 pints (pt)
- 1 liter (l) = 1.06 quarts (qt)
- 1 liter (l) = 0.26 gallon (gal)
- 1 cubic meter (m^3) = 36 cubic feet (cu ft, ft^3)
- 1 cubic meter (m^3) = 1.3 cubic yards (cu yd, yd^3)

TEMPERATURE (EXACT)
$[(9/5)\,y + 32]\,°C = x\,°F$

QUICK INCH - CENTIMETER LENGTH CONVERSION

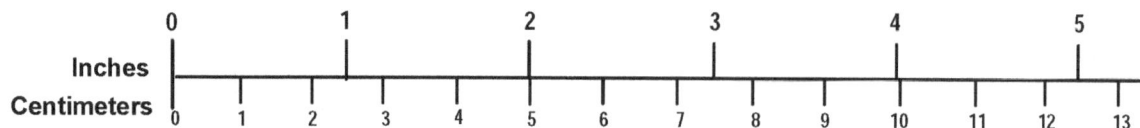

QUICK FAHRENHEIT - CELSIUS TEMPERATURE CONVERSION

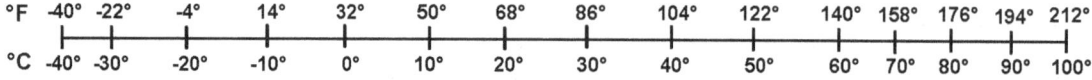

For more exact and or other conversion factors, see NIST Miscellaneous Publication 286, Units of Weights and Measures. Price $2.50 SD Catalog No. C13 10286

Updated 6/17/98

TABLE OF CONTENTS

EXECUTIVE SUMMARY vii
1. INTRODUCTION 1
2. OBJECTIVE TESTS 4
   2.1. Overview 4
   2.2. General Requirements 5
   2.3. Various Speed Approach Tests 6
      2.3.1. Traffic Signal Tests 7
      2.3.2. Stop Sign Tests 7
   2.4. Approach Lane Tests 10
      2.4.1. Edge of Approach Warning Tests 10
      2.4.2. Edge of Approach Tests for Nuisance Warning 12
   2.5. Late Lane Shift Tests 13
      2.5.1. Late Lane Shift Tests for Warning 13
      2.5.2. Late Lane Shift Tests for Nuisance Warning 14
   2.6. Multiple Intersection Tests (within 300-Meter Radius) 16
      2.6.1. Multiple Intersection Tests for Warning 16
      2.6.2. Multiple Intersection Tests for Nuisance Warning 19
   2.7. Dynamic Signal Change Tests 19
      2.7.1. Dynamic Signal Change to Yellow (Too Late to Warn) 20
      2.7.2. Dynamic Signal Change to Red (Sufficient to Warn) 21
      2.7.3. Dynamic Signal Change to Green (No Warning Necessary) 23
   2.8. Signal Phase and Timing Reception and Reflection Tests 26
   2.9. Objective Test Conclusions 27
3. PILOT TEST 29
   3.1. Overview 29
   3.2. Pseudo-Naturalistic Test Results 29
   3.3. Test Track Results 31
   3.4. Pilot Test Questionnaires 32
   3.5. Pilot Test Conclusions 33
4. OVERALL RECOMMENDATIONS FOR CICAS-V SYSTEM 35
5. REFERENCES 36

LIST OF TABLES

Table 1. Description of Objective Test Scenarios 3
Table 2. Test Results of Red Light and Stop Sign Approaches at Various Speeds 4
Table 3. Results of Edge of Approach Lane and Late Lane Shift Tests 4
Table 4. Multiple Intersection Test Results 4
Table 5. Dynamic Signal Change Test Results 5
Table 6. Engineering Test Results 5
Table 7. Results for Tests Runs – Approaches to Red Light at Various Speeds 8
Table 8. Results for Tests Runs – Approaches to a Stop Sign at Various Speeds 9
Table 9. Results for Tests Runs – Edge of Approach Warning Test 11
Table 10. Results for Tests Runs – Edge of Approach Warning Test 12
Table 11. Results for Late Lane Shift Warning Tests 15
Table 12. Results for Late Lane Shift Warning Tests 16
Table 13. Results for Multiple Intersection Warning Tests 18
Table 14. Results for Multiple Intersection Nuisance Warning Tests 19
Table 15. Results for Dynamic Signal Change to Yellow Tests 21
Table 16. Results for Dynamic Signal Change to Red (Sufficient to Warn) Tests 24
Table 17. Results for Dynamic Signal Change to Green (No Warning Necessary) Tests 25
Table 18. Results of SPaT Reflection and Reception Tests 28
Table 19. Distribution of Evaluated Drivers by Age and Gender 29
Table 20. Turn Maneuvers by Intersection Type in Pseudo-Naturalistic Study 29
Table 21. Distribution of "Stop-Controlled Algorithm 1" Drivers by Age and Gender ... 30
Table 22. Distribution of Sufficiently Distracted Test Track Drivers by Age and Gender 32
Table 23. Distribution of Questionnaire Respondents by Alerts Experienced 33

LIST OF FIGURES

Figure 1. Geometry of Tests for Warning at Various Approach Speeds 6

Figure 2. Geometry of Tests for Edge of Approach Lane Warning 10

Figure 3. Geometry of Tests for Edge of Approach Lane Nuisance Warning 12

Figure 4. Geometry of Tests for Late Lane Shift Warning 13

Figure 5. Geometry of Tests for Late Lane Shift Nuisance Warning 14

Figure 6. Geometry of Tests for Multiple Intersection Warning 17

Figure 7. Geometry for Dynamic Signal Change Warning Tests 20

Figure 8. Time Sequence Schematic for Dynamic Signal Change to Yellow Tests 21

Figure 9. Geometry for Dynamic Signal Change to Red (Sufficient to Warn) Tests 22

Figure 10. Time Sequence Schematic for Dynamic Signal Change to Red (Sufficient to Warn) Tests 22

Figure 11. Geometry for Dynamic Signal Change to Green (No Warning Necessary) Tests 23

Figure 12. Time Sequence Schematic for Dynamic Signal Change to Green (No Warning Necessary) Tests 25

Figure 13. Geometry for SPaT Reflection and Reception Tests 26

Figure 14. Map of Pseudo-Naturalistic Study Route with Labeled Intersections 30

LIST OF ACRONYMS

CAMP	Crash Avoidance Metrics Partnership
CICAS-V	Cooperative Intersection Collision Avoidance System for Violation
DAS	Data Acquisition System
DGPS	Differential Global Positioning System
DSRC	Dedicated Short Range Communication
FOT	Field Operational Test
GID	Geographical Intersection Description
GPS	Global Positioning System
OBE	On-Board Equipment
RSE	Roadside Equipment
SPaT	Signal Phase and Timing
TTI	Time-To-Intersection
VTTI	Virginia Tech Transportation Institute

EXECUTIVE SUMMARY

This report characterizes the capability of the Cooperative Intersection Collision Avoidance System for Violations (CICAS-V) based on data collected from objective tests and a pilot test conducted on the Smart Road test track and public roads in Blacksburg, VA. The CICAS-V is a prototype intersection violation warning system that alerts drivers who are about to run a red light or stop sign. Twelve types of objective tests were conducted to examine CICAS-V performance under defined conditions and operating parameters. Objective test scenarios were selected to evaluate the readiness and maturity of the system to issue timely warnings in potential red light or stop sign violations and to suppress alerts in situations when there is no risk of violation. The pilot test gave nearly one hundred naive drivers the opportunity to experience the CICAS-V on public roads. They each drove one of two CICAS-V-equipped vehicles unaccompanied on a prescribed two-hour course that made a total of 52 crossings through 13 system-equipped intersections. A subset of the subjects participated in a follow-on test that was designed to induce an unexpected warning on the Smart Road test track.

Results of the objective tests show that the CICAS-V behaves as designed in a wide variety of common driving situations. The test vehicle consistently warned the driver when the vehicle was exceeding the target speed for a safe stop in a lane designated to stop, whether by a stop sign or by a traffic signal, over a large range of test speeds. The system consistently distinguished between the required alarm state for the current lane and that of nearby lanes and was sufficiently robust even if the vehicle were located at the edge of the designated lane or dynamically shifted between lanes in which the appropriate alert status changed. It also differentiated between multiple intersections in close proximity and engaged a warning state appropriate for the relevant intersection and lane. The CICAS-V was very robust in its ability to evaluate the situation and warn correctly under conditions that severely inhibited the line-of-sight wireless reception.

Results of the pilot test confirm that the CICAS-V functions as implemented on public roads. The system reacted appropriately in the vast majority of the 2,618 stop controlled intersection crossings and the 1,455 signal controlled intersection crossings recorded during the test. Some nuisance warnings were issued due to a flaw in the stop-controlled warning, which were immediately remedied by a minor fix. Invalid signal-controlled warnings were also observed due to an erroneous geographical intersection description that was quickly corrected. The CICAS-V appropriately warned three drivers who may have inadvertently violated an intersection controlled by a partially obscured stop sign and one driver who might have otherwise violated a red traffic signal. It also appropriately warned all 18 intentionally-distracted drivers on the test track, facilitating seventeen of them to avoid a violation.

The CICAS-V tests demonstrate the practical need for the fine-tuning of a prototype system before commencing a full-scale field operational test, particularly in the actual region where the field test will be conducted.

1. INTRODUCTION

This report presents the results of the objective test and pilot test tasks of the Cooperative Intersection Collision Avoidance System limited to Stop Sign and Traffic Signal Violations (CICAS-V) project. The system was designed by the Crash Avoidance Metrics Partnership (CAMP), a consortium of automotive original equipment manufacturers. The road-side equipment (RSE) component of the system at any intersection broadcasts a Geographical Intersection Description (GID), local corrections for a Global Positioning System (GPS), and the current state of any signals in the intersection. The on-board equipment (OBE) then uses data from the vehicle and its own GPS to assess the vehicle's lane of approach and ability to stop safely (if required) without excessive deceleration. If its algorithm determines a warning is necessary to prevent a violation of the intersection, it issues visual, audio, and haptic alerts. The haptic alert is in the form of a 600-millisecond brake pulse.

The objective tests were conducted on the Smart Road test track at the Virginia Tech Transportation Institute (VTTI) in Blacksburg, VA, between July 15 and July 17, 2008. The pilot test took place on public roads in the Blacksburg, VA area and on the Smart Road test track. The test vehicles and the roadside equipment were equipped with data acquisition systems capable of recording relevant test parameters for quantitative analysis after the test. Additional data (e.g., the perception of each alarm component) were explicitly recorded by test personnel. Specifics of the test procedures are given in two CAMP project reports [1, 2].

The goal of the CICAS-V program is to assess the safety benefits, driver acceptance, and system capability associated with a prototype intersection violation warning system. The first phase of the project developed and tested the CICAS-V. Based on the successful completion of the first phase, a fleet of CICAS-V vehicles is to be built and tested in a Field Operational Test (FOT). The purpose of this document is to review the CICAS-V performance data from the first phase of development to determine if the system is ready for FOT deployment.

Objective tests were conducted to examine CICAS-V performance under defined conditions and operating parameters. Objective test scenarios were selected to evaluate the readiness and maturity of the system to issue timely warnings in potential red light or stop sign violations and to suppress alerts in situations when there is no risk of running the red light or stop sign. The objective tests were divided into twelve types:

1. Red Light Approaches at Various Speeds (Warning)
2. Stop Sign Approaches at Various Speeds (Warning)
3. Edge of Approach Test (Warning)
4. Edge of Approach Test (Nuisance)
5. Late Lane Shift Test (Warning)
6. Late Lane Shift Test (Nuisance)
7. Multiple Intersections - 300 m Radius (Warning)

8. Multiple Intersections - 300 m Radius (Nuisance)
9. Dynamic Signal Change to Yellow (Too Late to Warn)
10. Dynamic Signal Change to Red (Sufficient to Warn)
11. Dynamic Signal Change to Green (No Warning)
12. Signal Phase and Timing (SPaT) Reflection and Reception

Table 1 lists the objective test scenarios and their characteristics, including warning tests, nuisance tests, and engineering tests. In the nuisance (no warning) tests, common vehicle maneuvers are performed to demonstrate that the warning system does not issue alerts that could be considered a nuisance by the driver. Engineering tests exercised the limits of system performance under unfavorable conditions.

Initial conditions are used to define a test run and judge whether it was conducted correctly; i.e., that it was a "valid" run. The pass/fail criteria for each test are discussed in the appropriate sections of this report. All warning and nuisance tests were conducted and evaluated by applying the respective run validity and pass/fail criteria for the given test. Engineering tests were conducted in a similar manner to the required tests, although they simply measured the tolerance of the system to adverse conditions and were not pass/fail tests per se.

The pilot test for the field operational test gave nearly one hundred naive drivers the opportunity to experience the CICAS-V system on public roads. They each drove one of two CICAS-V-equipped vehicles unaccompanied on a prescribed two-hour course that made a total of 52 crossings through 13 system-equipped intersections (ten stop sign-controlled intersections and three signal-controlled intersections). A subset of the subjects participated in a follow-on test that was designed to induce an unexpected warning on the Smart Road test track.

Table 1. Description of Objective Test Scenarios

Scenario	Speed	Signal / Stop Sign	SPaT Used	Test Type
Red Light Approaches at Various Speeds	55 ±2.5 mph 35 ±2.5 mph 25 ±2.5 mph	Signal	Fixed Red	Warning
Stop Sign Approaches at Various Speeds	25 ±2.5 mph 35 ±2.5 mph 55 ±2.5 mph	Stop Sign	No SPaT – Stop Sign Only	Warning
Edge of Approach Test	35 ±2.5 mph	Signal	Fixed Red and Green	Warning
End of Approach Test	35 ±2.5 mph	Signal	Fixed Red and Green	Nuisance
Late Lane Shift Test	35 ±2.5 mph	Signal	Fixed Red and Green	Warning
Late Lane Shift Test	35 ±2.5 mph	Signal	Fixed Red and Green	Nuisance
Multiple Intersections - 300 m Radius	35 ±2.5 mph	Signal	Fixed Red for All Approaches, All Intersections	Warning
Multiple Intersections - 300 m Radius	35 ±2.5 mph	Signal	Fixed Green for Main, Fixed Red for Alternate	Nuisance
Dynamic Signal Change to Yellow (Too Late to Warn)	35 ±2.5 mph	Signal	Dynamic – Changes Triggered by Vehicle Position	Nuisance
Dynamic Signal Change to Red (Sufficient to Warn)	35 ±2.5 mph	Signal	Dynamic – Changes Triggered by Vehicle Position	Warning
Dynamic Signal Change to Green	35 ±2.5 mph	Signal	Dynamic – Changes Triggered by Vehicle Position	Nuisance
SPaT Reflection and Reception	35 ±5.0 mph	Signal	Fixed Red and Green	Engineering

1 mph = 1.61 km/h

2. OBJECTIVE TESTS

2.1. Overview

The objective tests described in this report were run and evaluated by applying the respective run validity and pass/fail criteria for the given test. A run was considered valid if the analysis of the data acquisition system (DAS) data met the run validity requirement for the given set of tests. Pass/fail criteria were only applied to valid runs, and pass/fail determinations were made based on measurements recorded in the DAS.

Tables 2 through 6 summarize the results for five groups of test scenarios. Each table presents the pass/fail results of each test category including the number of runs, number of valid runs based on the test initial condition requirements, and number of passed runs.

Table 2. Test Results of Red Light and Stop Sign Approaches at Various Speeds

Speed [mph]		Signal/Stop Sign	Runs	Valid Runs	Pass	Fail	Result
Nominal	Actual Average						
25	25.4	Signal	8	8	8	0	Pass
35	35.0	Signal	9	9	9	0	Pass
55	54.5	Signal	9	9	8	1*	Pass
25	24.6	Stop Sign	8	8	8	0	Pass
35	34.7	Stop Sign	8	8	8	0	Pass
55	53.9	Stop Sign	8	8	8	0	Pass

*Failure due to lack of haptic warning only

Table 3. Results of Edge of Approach Lane and Late Lane Shift Tests

Scenario	Runs	Valid Runs	Pass	Fail	Result
Edge of Approach Lane (Warning)	16	16	16	0	Pass
Edge of Approach Lane (Nuisance)	16	15	15	0	Pass
Late Lane Shift (Warning)	10	9	9	0	Pass
Late Lane Shift (Nuisance)	8	8	8	0	Pass

Table 4. Multiple Intersection Test Results

Scenario	Runs	Valid Runs	Pass	Fail	Result
Multiple Intersections – 300 m Radius (Warning)	12	6	6	0	Pass
Multiple Intersections – 300 m Radius (Nuisance)	12	5	5	0	Pass

Table 5. Dynamic Signal Change Test Results

Scenario	Runs	Valid Runs	Pass	Fail	Result
Dynamic Signal Change to Yellow (Too Late to Warn)	10	10	10	0	Pass
Dynamic Signal Change to Red (Sufficient to Warn)	10	10	10	0	Pass
Dynamic Signal Change to Green (No Warning)	8	8	8	0	Pass

Table 6. Engineering Test Results

Scenario	Runs	Valid Runs	Pass	Fail	Result
Following Tractor Trailer by Less than 6 meters	8	8	8	0	Pass

2.2. General Requirements

Each set of tests was designed to assess the system's ability to perform appropriately under defined conditions. Test data from the DAS were recorded in a well-defined format for further analysis. If the conditions of the test were met (e.g., the vehicle speed is within the prescribed limits), the test was declared valid. If the system performed as expected (e.g., no warning or all warnings within the prescribed interval, as appropriate) in a valid run, the system was deemed to have passed that run of the test. If the system responded inappropriately (e.g., a warning when none was required, a warning outside the appropriate range, an appropriate warning that did not occur), the system was judged to have failed for that run. The system was judged to have passed a test series if it passed at least six of eight valid runs. As a result of stricter validation criteria than the researchers on the track were using, some test sets have been evaluated with fewer than eight valid runs. Invalid runs were not used to evaluate the system.

The general run validity requirements include GPS coverage and initial vehicle speed:

- For minimum valid GPS coverage, the data acquisition system must confirm that:
  - The standard deviation of the estimated GPS position was no more than 1.5 meters in the horizontal plane,
  - The Position Dilution of Precision [a dimensionless parameter for assessment of precision based on the geometry of the constellation of satellites in use] was no more than 5.0, and
  - There were at least five satellites used in the calculation of GPS position.
- The vehicle speed must be within 2.5 mph (4 km/h) of the nominal value for the test to be considered valid.

Only one test was declared invalid for violating these general requirements (due to an error ellipse dimension of 1.6 meters).

There were additional validity requirements for some test series. Some of these were lane change timing, signal change timing, lane edge distance, and the following distance behind a lead vehicle. The appropriate geographic lane configuration, travel lane, and signal phase were explicitly verified for each test run. It was discovered during testing that the GID was displaced approximately 0.5 m laterally and 0.5 m longitudinally from the physical lane markings, which was important to note when evaluating the results of certain sets of tests.

2.3. Various Speed Approach Tests

Results are presented below for scenarios that involve the subject vehicle approaching an intersection at various speeds in which the vehicle would be required to stop for either a red signal light or a stop sign. The algorithm for calculating the optimum timing of a warning as a function of the vehicle's actual speed was determined in a previous task of the CICAS-V project [3]. Any warning was expected to occur within 200 msec (0.200 sec) of this optimum point. Although the test track was marked with the location of where this region would begin and end if the vehicle traveled at nominal speed, quantitative evaluation was based on the recorded test speed (rounded to the nearest full km/h) and warning location. The warning distances were prescribed in appendix tables (Table 641-11 for signaled intersections, Table 741-9 for stop-controlled intersections) in the CAMP report [4].

The overall geometry of this test is shown in Figure 1. The optimum location for a warning for a vehicle traveling at the nominal speed would be in the green band. A warning in either red section would constitute a failure for the test run.

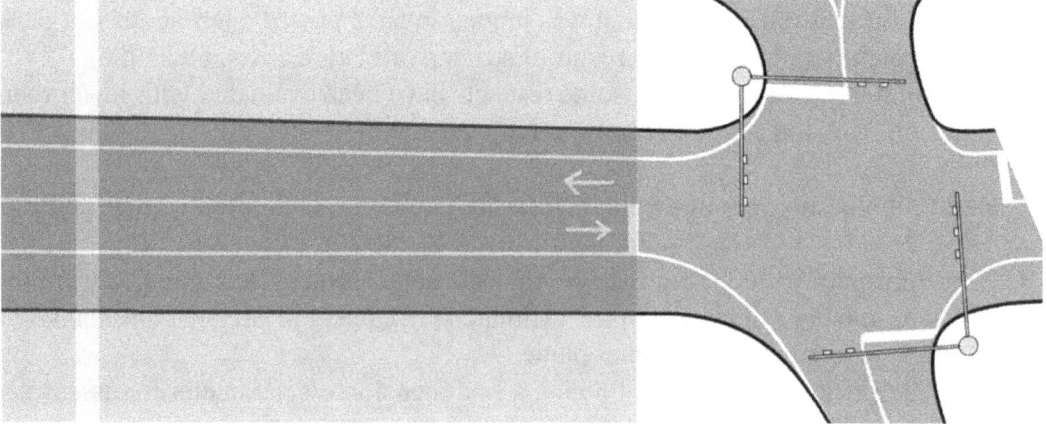

Figure 1. Geometry of Tests for Warning at Various Approach Speeds

2.3.1. Traffic Signal Tests

In this test, the traffic signal in the intersection was red. Eight runs were conducted at a nominal speed of 25 mph (40.2 km/h). Nine each were also conducted at two other nominal speeds: 35 mph (56.3 km/h) and 55 mph (88.5 km/h). The test vehicle was driven toward the intersection in a valid approach lane until a warning was received. No invalid runs occurred.

Table 7 summarizes the results for this series of tests. The pass/fail criteria for these tests required an audio alert, a visual icon, and a haptic warning (a 600 msec pulse of the brakes) at a prescribed distance (plus or minus tolerance) from the intersection. The tolerance was computed as the distance traveled at the actual test speed in 0.200 sec.

The system failed in only one run. In the fourth run at 55 mph, the haptic warning did not occur. The other two warnings (audio alert and visual icon) were issued. Nonetheless, this incomplete warning resulted in a failed test run, but not a failure for the set of tests neither at the 55 mph nominal speed nor in the group of traffic signal tests at various speeds. The average deviation from the nominal warning distance was +0.2 m with a standard deviation of 0.6 m for the 25 mph tests, +0.3 m with a standard deviation of 0.4 m for the 35 mph tests, and +0.4 m with a standard deviation of 1.8 m for the 55 mph tests.

2.3.2. Stop Sign Tests

In this test, a stop sign was located at the test intersection. Eight runs were conducted at each of three nominal speeds: 25 mph, 35 mph, and 55 mph. The test vehicle was driven toward the intersection in a valid approach lane until a warning was received. No invalid runs occurred. The pass/fail criteria for these tests also required an audio alert, a visual icon, and a haptic warning at a prescribed distance (plus or minus tolerance) from the intersection. The tolerance was computed as the distance traveled at the actual test speed in 0.200 sec.

Table 8 summarizes the results for this series of tests. The system passed each test run at each speed. The average deviation from the nominal warning distance was +0.1 m with a standard deviation of 0.5 m for the 25 mph tests, +0.5 m with a standard deviation of 1.2 m for the 35 mph tests, and +0.1 m with a standard deviation of 1.0 m for the 55 mph tests.

Table 7. Results for Tests Runs – Approaches to Red Light at Various Speeds

Set	Run	Nominal Test Speed [mph]	Recorded Test Speed [mph]	Nominal Warning Distance [m]	Actual Warning Distance [m]	Minimum Warning Distance [m]	Maximum Warning Distance [m]	Warning Distance Deviation [m]	Warning Within Specified Range?	All Warnings Observed?	Test Conditions Valid?	Pass/Fail
1	1	55	54.4	98.3	100.8	93.3	103.2	2.5	Yes	Yes	Yes	Pass
	2	55	54.5	98.3	97.9	93.4	103.1	-0.4	Yes	Yes	Yes	Pass
	3	55	54.3	98.3	98.5	93.4	103.1	0.2	Yes	Yes	Yes	Pass
	4	55	54.7	98.3	101.3	93.4	103.1	3	Yes	No	Yes	Fail
	5	55	53.6	96.0	95.3	91.2	100.8	-0.7	Yes	Yes	Yes	Pass
	6	55	55.4	102.9	103.5	97.9	107.8	0.6	Yes	Yes	Yes	Pass
	7	55	54.4	98.3	99.7	93.4	103.2	1.4	Yes	Yes	Yes	Pass
	8	55	54.8	100.6	97.7	95.7	105.5	-2.9	Yes	Yes	Yes	Pass
	9	55	54.4	98.3	98.1	93.4	103.2	-0.2	Yes	Yes	Yes	Pass
2	1	35	37.0	44.7	45.4	41.4	48.0	0.7	Yes	Yes	Yes	Pass
	2	35	35.1	40.2	40.2	37.1	43.4	0.0	Yes	Yes	Yes	Pass
	3	35	34.6	38.8	38.8	35.7	41.9	0.0	Yes	Yes	Yes	Pass
	4	35	34.7	38.8	40.0	35.7	41.9	1.2	Yes	Yes	Yes	Pass
	5	35	34.8	40.2	40.3	37.1	43.3	0.1	Yes	Yes	Yes	Pass
	6	35	35.0	40.2	40.5	37.1	43.3	0.3	Yes	Yes	Yes	Pass
	7	35	34.2	38.8	38.8	35.7	41.8	0.0	Yes	Yes	Yes	Pass
	8	35	34.8	40.2	40.6	37.1	43.3	0.4	Yes	Yes	Yes	Pass
	9	35	35.1	40.2	40.5	37.1	43.3	0.3	Yes	Yes	Yes	Pass
3	1	25	25.1	20.2	20.0	17.9	22.4	-0.2	Yes	Yes	Yes	Pass
	2	25	24.9	20.2	20.1	18.0	22.4	-0.1	Yes	Yes	Yes	Pass
	3	25	24.1	18.2	19.3	16.0	20.3	1.1	Yes	Yes	Yes	Pass
	4	25	24.2	19.2	19.4	17.0	21.3	0.2	Yes	Yes	Yes	Pass
	5	25	24.4	19.2	18.8	17.0	21.3	-0.4	Yes	Yes	Yes	Pass
	6	25	24.1	18.2	19.5	16.0	20.3	1.3	Yes	Yes	Yes	Pass
	7	25	24.3	19.2	18.9	17.0	21.3	-0.3	Yes	Yes	Yes	Pass
	8	25	24.4	19.2	19.2	17.0	21.3	0.0	Yes	Yes	Yes	Pass

Table 8. Results for Tests Runs – Approaches to a Stop Sign at Various Speeds

Set	Run	Nominal Test Speed [mph]	Recorded Test Speed [mph]	Nominal Warning Distance [m]	Actual Warning Distance [m]	Minimum Warning Distance [m]	Maximum Warning Distance [m]	Warning Distance Deviation [m]	Warning Within Specified Range?	All Warnings Observed?	Test Conditions Valid?	Pass / Fail
1	1	35	34.2	33.5	33.2	30.4	36.6	-0.3	Yes	Yes	Yes	Pass
1	2	35	34.5	33.5	34.8	30.4	36.6	1.3	Yes	Yes	Yes	Pass
1	3	35	35.6	36.8	36.1	33.6	40.0	-0.7	Yes	Yes	Yes	Pass
1	4	35	34.6	33.5	36.3	30.4	36.6	2.8	Yes	Yes	Yes	Pass
1	5	35	33.8	31.9	31.2	28.9	34.9	-0.7	Yes	Yes	Yes	Pass
1	6	35	35.2	35.1	35.1	32.0	38.2	0.0	Yes	Yes	Yes	Pass
1	7	35	35.1	35.1	35.8	32.0	38.2	0.7	Yes	Yes	Yes	Pass
1	8	35	34.9	35.1	35.7	32.0	38.2	0.6	Yes	Yes	Yes	Pass
2	1	25	24.7	13.9	13.9	11.7	16.1	0.0	Yes	Yes	Yes	Pass
2	2	25	24.7	13.9	14.8	11.7	16.1	0.9	Yes	Yes	Yes	Pass
2	3	25	24.3	13.9	13.7	11.7	16.1	-0.2	Yes	Yes	Yes	Pass
2	4	25	24.3	13.9	13.4	11.7	16.1	-0.5	Yes	Yes	Yes	Pass
2	5	25	25.2	14.8	15.2	12.6	17.1	0.4	Yes	Yes	Yes	Pass
2	6	25	24.9	14.8	15.2	12.6	17.0	0.4	Yes	Yes	Yes	Pass
2	7	25	24.6	13.9	14.1	11.7	16.1	0.2	Yes	Yes	Yes	Pass
2	8	25	24.1	13.1	12.6	10.9	15.2	-0.5	Yes	Yes	Yes	Pass
3	1	55	53.9	110.1	109.8	105.3	114.9	-0.3	Yes	Yes	Yes	Pass
3	2	55	53.7	110.1	111.1	105.3	114.9	1.0	Yes	Yes	Yes	Pass
3	3	55	53.9	110.1	109.6	105.3	114.9	-0.5	Yes	Yes	Yes	Pass
3	4	55	53.7	110.1	110.0	105.3	114.9	-0.1	Yes	Yes	Yes	Pass
3	5	55	54.4	113.6	113.0	108.7	118.4	-0.6	Yes	Yes	Yes	Pass
3	6	55	53.3	106.7	108.9	101.9	111.4	2.2	Yes	Yes	Yes	Pass
3	7	55	54.0	110.1	109.8	105.2	114.9	-0.3	Yes	Yes	Yes	Pass
3	8	55	54.4	113.6	112.8	108.7	118.4	-0.8	Yes	Yes	Yes	Pass

2.4. Approach Lane Tests

In signaled intersections, different approach lanes may be associated with different path intentions (e.g., a left turn only lane) and thus may require different warning strategies. The approach lane tests seek to verify that the GPS and Differential GPS (DGPS) systems can work in conjunction with the GID to accurately determine in which lane the vehicle is moving and adjust its warnings appropriately. Recall that the GID on the test track was displaced about 0.5 m laterally from the physical lane markings. A visual confirmation that the right wheels were steadily within 0.5 m of the right side lane marker could therefore be interpreted by the system as the wheels being up to 0.5 m over the lane marker. Thus, these sets of tests inadvertently tested the robustness of the system to deficient GIDs.

2.4.1. Edge of Approach Warning Tests

This test seeks to verify that the system will appropriately warn the driver even if the vehicle is near the edge of the approach lane. In this case, for any run to be valid, the right tires must be within 0.5 m (20 inches) of the right lane marker. That is, they must be 0.25 ± 0.25 m to the left of the right lane boundary. The SPaT system transmitted a red light signal for the specified approach lane and green light signals for the lanes on either side. The geometry of the test setup is shown in Figure 2. The green zone indicates the approximate region in which a warning would be appropriate at the nominal speed of 35 mph.

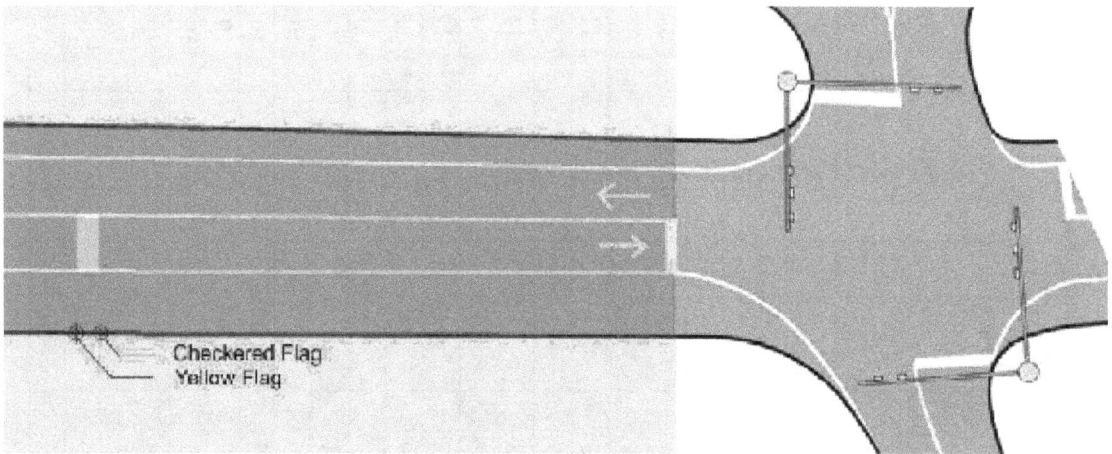

Figure 2. Geometry of Tests for Edge of Approach Lane Warning

Sixteen test runs were conducted. All of the runs were valid. The pass/fail criteria were the same as those in Section 2.3.1 – all three warning modes within a 0.200 sec tolerance of the warning distance from Table 641-11 [4]. All 16 runs passed. The results for each run are given in Table 9. The average deviation from the nominal warning distance was +0.6 m with a standard deviation of 0.4 m.

Table 9. Results for Tests Runs – Edge of Approach Warning Test

Run	Nominal Test Speed [mph]	Recorded Test Speed [mph]	Nominal Warning Distance [m]	Actual Warning Distance [m]	Minimum Warning Distance [m]	Maximum Warning Distance [m]	Warning Distance Deviation [m]	Warning Within Specified Range?	All Warnings Observed?	Test Conditions Valid?	Pass / Fail
1	35	34.5	38.8	39.7	35.7	41.8	0.9	Yes	Yes	Yes	Pass
2	35	34.7	38.8	39.3	35.7	41.8	0.5	Yes	Yes	Yes	Pass
3	35	34.5	38.8	39.6	35.7	41.8	0.8	Yes	Yes	Yes	Pass
4	35	34.6	38.8	39.3	35.7	41.8	0.5	Yes	Yes	Yes	Pass
5	35	34.6	38.8	39.1	35.7	41.8	0.3	Yes	Yes	Yes	Pass
6	35	34.5	38.8	40.0	35.7	41.8	1.2	Yes	Yes	Yes	Pass
7	35	34.4	38.8	39.9	35.7	41.8	1.1	Yes	Yes	Yes	Pass
8	35	34.6	38.8	39.4	35.7	41.8	0.6	Yes	Yes	Yes	Pass
9	35	34.5	38.8	39.4	35.7	41.8	0.6	Yes	Yes	Yes	Pass
10	35	34.6	38.8	38.8	35.7	41.8	0.0	Yes	Yes	Yes	Pass
11	35	34.4	38.8	38.9	35.7	41.8	0.1	Yes	Yes	Yes	Pass
12	35	34.6	38.8	38.9	35.7	41.8	0.1	Yes	Yes	Yes	Pass
13	35	34.6	38.8	39.1	35.7	41.8	0.3	Yes	Yes	Yes	Pass
14	35	34.5	38.8	39.3	35.7	41.8	0.5	Yes	Yes	Yes	Pass
15	35	34.6	38.8	39.7	35.7	41.8	0.9	Yes	Yes	Yes	Pass
16	35	34.6	38.8	39.7	35.7	41.8	0.9	Yes	Yes	Yes	Pass

2.4.2. Edge of Approach Tests for Nuisance Warning

This set of edge of approach tests was identical to the previous set except that the specified approach lane was the only one with a green rather than a red signal light. The test sought to confirm that the system would not give an erroneous warning when no warning is required even though the vehicle is traveling within 0.5 m of a lane in which a warning would be appropriate. Figure 3 illustrates this scenario.

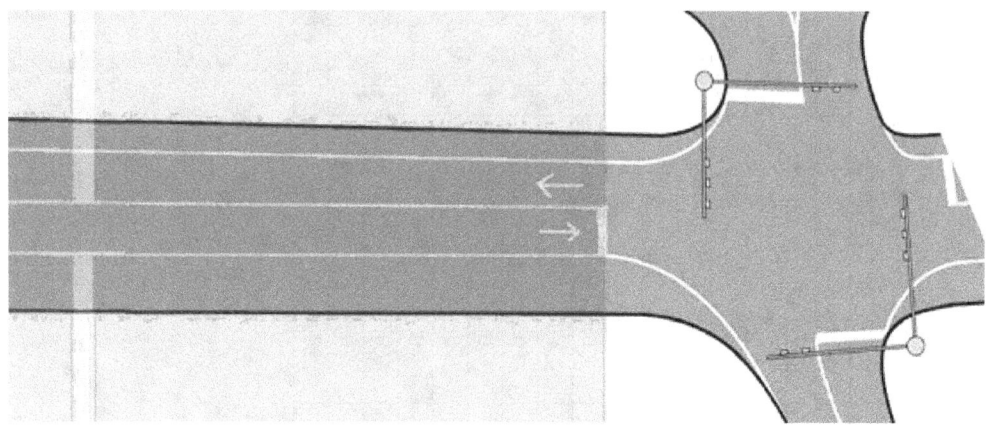

Figure 3. Geometry of Tests for Edge of Approach Lane Nuisance Warning

Once again, sixteen runs were conducted at a nominal speed of 35 mph. Only one of the sixteen runs was not valid using the stricter validation conditions imposed by the evaluators. The pass/fail criterion was that no warning should be observed. All fifteen valid runs passed as seen in Table 10.

Table 10. Results for Tests Runs – Edge of Approach Warning Test

Run	Nominal Test Speed [mph]	Speed Within Specified Range?	Lane Position Within Specified Range?	Test Conditions Valid?	Any Warning Observed?	Pass/ Fail
1	35	Yes	Yes	Yes	No	Pass
2	35	Yes	Yes	Yes	No	Pass
3	35	Yes	Yes	Yes	No	Pass
4	35	Yes	Yes	Yes	No	Pass
5	35	Yes	No	No		
6	35	Yes	Yes	Yes	No	Pass
7	35	Yes	Yes	Yes	No	Pass
8	35	Yes	Yes	Yes	No	Pass
9	35	Yes	Yes	Yes	No	Pass
10	35	Yes	Yes	Yes	No	Pass
11	35	Yes	Yes	Yes	No	Pass
12	35	Yes	Yes	Yes	No	Pass
13	35	Yes	Yes	Yes	No	Pass
14	35	Yes	Yes	Yes	No	Pass
15	35	Yes	Yes	Yes	No	Pass
16	35	Yes	Yes	Yes	No	Pass

2.5. Late Lane Shift Tests

In signaled intersections, different approach lanes may be associated with different path intentions (e.g., a left-turn only lane). When a vehicle dynamically shifts from one lane to another when approaching an intersection, the CICAS-V system must be able to accurately determine in which lane the vehicle is traveling and whether or not it is appropriate to warn. These tests sought to verify not only that lane position was accurate, but also that it was in a timely manner to limit inappropriate responses.

2.5.1. Late Lane Shift Tests for Warning

This test sought to verify that the CICAS-V system would appropriately warn the driver when the vehicle shifts into a lane requiring a warning, even if the vehicle has not completed the process of changing lanes at the optimum warning distance.

In this test, the vehicle travels at a nominal speed of 35 mph. Two flags denoting the start and ending location for the lane change maneuver are located such that, at nominal speed, the vehicle will start the maneuver 1.5 seconds before reaching the optimum warning location (as defined by Table 641-11 [4]) and finish no later than 2.5 seconds after reaching the optimum warning distance. It was assumed that the maneuver would take no longer than four seconds. A test was considered valid if the lane change started after 1.5 seconds before the optimum warning distance and finished before 2.5 seconds after the optimum warning distance. The geometry for this test is shown in Figure 4.

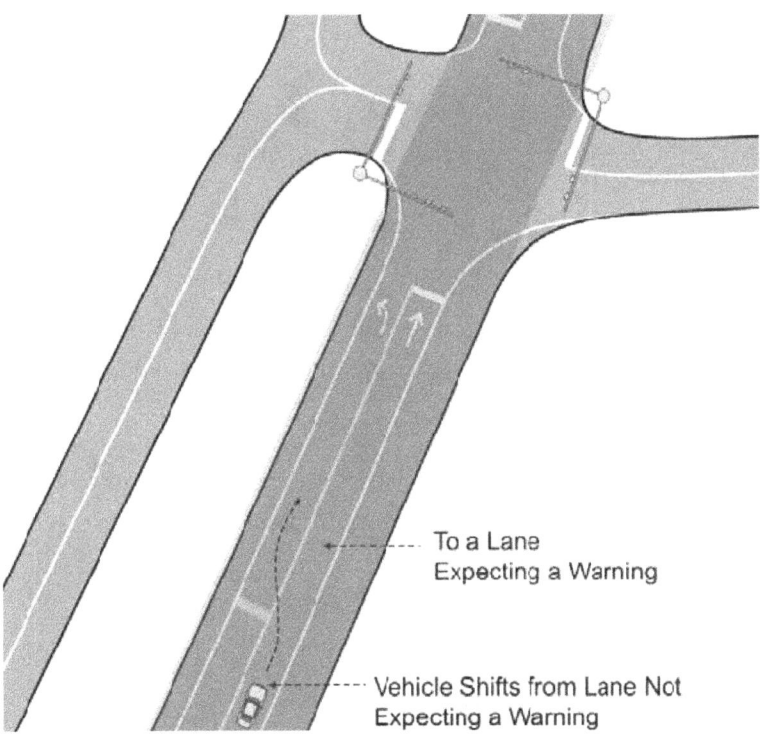

Figure 4. Geometry of Tests for Late Lane Shift Warning

Ten tests of this type were run. In the first run, the lane shift maneuver was not completed within the allotted four seconds. The evaluators chose to categorize this run as invalid. Nonetheless, all three warnings did activate in this run, though after the vehicle had passed the minimum warning distance.

The pass/fail criterion for this test was that all three warnings would occur after the vehicle was in the lane expecting the warning. In the nine valid runs, all the warnings did occur within the nominal warning range, though all after the vehicle had passed the optimum warning location. Thus, all nine valid runs passed. The deviation of the warning distance in the nine valid runs was -2.0 m with a standard deviation of 0.4 m. Details of the runs are given in Table 11.

2.5.2. Late Lane Shift Tests for Nuisance Warning

This test sought to verify that a late lane shift from a lane requiring a warning to a lane not requiring a warning would not result in a false positive "nuisance" warning, even if the vehicle were still in the process of changing lanes as it passed what would be the optimum warning range (as defined by Table 641-11 [4]).

In this test, the vehicle traveled at a nominal speed of 35 mph. Two flags denoting the start and ending location for the lane change maneuver were located such that, at nominal speed, the vehicle would start the maneuver 1.5 seconds before reaching the optimum warning location (for the warning lane) and finish no later than 2.5 seconds after reaching the optimum warning distance. It is assumed that the maneuver will take no longer than four seconds. The geometry for this test is shown in Figure 5.

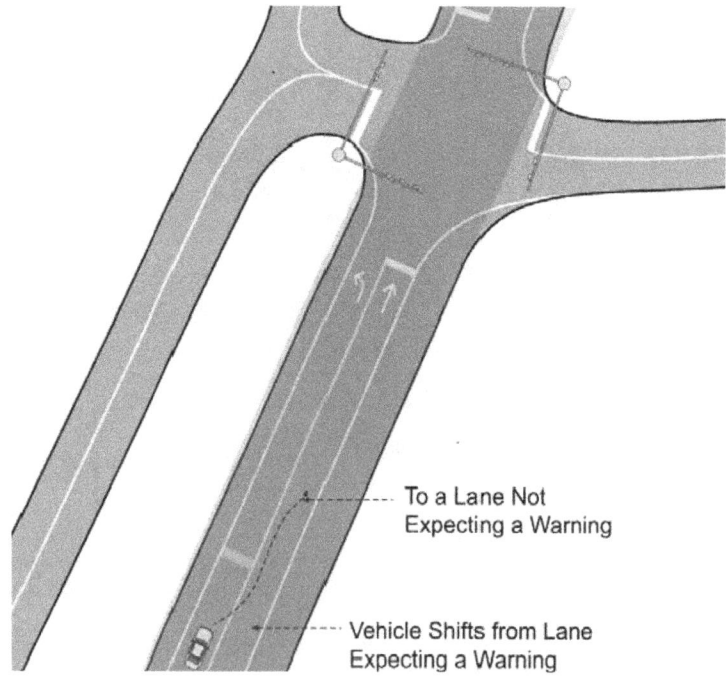

Figure 5. Geometry of Tests for Late Lane Shift Nuisance Warning

Table 11. Results for Late Lane Shift Warning Tests

Run	Nominal Test Speed [mph]	Recorded Test Speed [mph]	Nominal Warning Distance [m]	Actual Warning Distance [m]	Minimum Warning Distance [m]	Maximum Warning Distance [m]	Warning Distance Deviation [m]	Lane Change Within Four Seconds?	All Warnings Observed?	Test Conditions Valid?	Pass/Fail
1	35	34.6	38.8	34.6	35.7	41.8	-4.2	No	Yes	No	
2	35	34.5	38.8	37.7	35.7	41.8	-1.1	Yes	Yes	Yes	Pass
3	35	34.7	38.8	36.2	35.7	41.8	-2.6	Yes	Yes	Yes	Pass
4	35	34.7	38.8	36.8	35.7	41.8	-2.0	Yes	Yes	Yes	Pass
5	35	34.7	38.8	37.3	35.7	41.8	-1.5	Yes	Yes	Yes	Pass
6	35	34.7	38.8	36.7	35.7	41.8	-2.1	Yes	Yes	Yes	Pass
7	35	34.7	38.8	36.7	35.7	41.8	-2.1	Yes	Yes	Yes	Pass
8	35	34.5	38.8	36.7	35.7	41.8	-2.1	Yes	Yes	Yes	Pass
9	35	34.6	38.8	36.8	35.7	41.8	-2.0	Yes	Yes	Yes	Pass
10	35	34.4	38.8	36.7	35.7	41.8	-2.1	Yes	Yes	Yes	Pass

Eight tests of this type were run. In each run, the lane shift maneuver was completed within the allotted four seconds and in the allotted space. Thus, each run was valid. The pass/fail criterion was that no warning should occur. In the eight valid runs, no warnings occurred. Thus, all eight valid runs passed. Details of the runs are given in Table 12.

Table 12. Results for Late Lane Shift Warning Tests

Run	Nominal Test Speed [mph]	Speed Within Specified Range?	Lane Change Within Four Seconds?	Test Conditions Valid?	Any Warning Observed?	Pass / Fail
1	35	Yes	Yes	Yes	No	Pass
2	35	Yes	Yes	Yes	No	Pass
3	35	Yes	Yes	Yes	No	Pass
4	35	Yes	Yes	Yes	No	Pass
5	35	Yes	Yes	Yes	No	Pass
6	35	Yes	Yes	Yes	No	Pass
7	35	Yes	Yes	Yes	No	Pass
8	35	Yes	Yes	Yes	No	Pass

2.6. Multiple Intersection Tests (within 300-Meter Radius)

The CICAS-V system must be able to differentiate among multiple signaled intersections located in close proximity. On the VTTI Smart Road test track, the objective tests investigated the system's ability to handle two signaled intersections within 300 meters of each other.

2.6.1. Multiple Intersection Tests for Warning

The test track was set up with the GID and SPaT server indicating a second (virtual) intersection further down the road. Although there was no physical cross street, the GID was configured as if there were one, as shown in Figure 6. The second intersection was referred to as the "alternate intersection". Care was taken that the GID of the alternate intersection did not overlap with the GID of the main intersection.

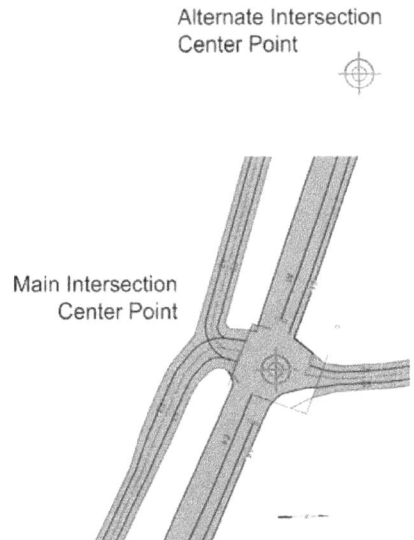

Figure 6. Geometry of Tests for Multiple Intersection Warning

For the warning test, the SPaT broadcasts that the signals were red for all approaches of both the main and the alternate intersections. The test was set up so that the warning was issued in the appropriate range for the closer intersection. That is, the test sought to verify that the CICAS-V system did not delay the appropriate warning for the main intersection simply because of an as-yet non-critical warning further down the expected path at the alternate intersection. This ability to identify and differentiate the GID of the two intersections will be crucial to the ability of the system to operate properly in a field operational test.

Beyond the normal test procedures and verifications of warning modalities, a test observer used a personal computer capable of Dedicated Short Range Communication (DSRC) reception and positioned before the earliest warning location to monitor the percentage of broadcast SPaT packets that arrived from the two intersections. The SPaT packets were recorded for the time period when the vehicle was within 300 m of the stop bar in a data file. The number of packets from the both intersections had to comprise at least 15 percent of the total number received for the run to be declared valid. This criterion was more stringent than that used at the test site. Twelve runs were conducted. In four runs, the number of packets from the alternate intersection was insufficient. In two cases, the number of packets from the main intersection was insufficient. These six runs were declared invalid.

Table 13 shows the results of this set of tests. The pass/fail criterion was that all warnings occur within the tolerance (actual test speed times 0.200 sec) of the optimal warning distance specified for the actual test speed in Table 641-11 [4]. In all six valid tests, all warnings were received within the appropriate range. Thus, all six valid tests passed. The average warning distance deviation in the valid runs was 1.2 m with a standard deviation of 1.0 m.

Table 13. Results for Multiple Intersection Warning Tests

Run	Nominal Test Speed [mph]	Recorded Test Speed [mph]	Nominal Warning Distance [m]	Actual Warning Distance [m]	Minimum Warning Distance [m]	Maximum Warning Distance [m]	Warning Distance Deviation [m]	Warning Within Specified Range?	All Warnings Observed?	Test Conditions Valid*?	Pass / Fail
1	35	34.5	38.8	39.3	35.7	41.8	0.5	Yes	Yes	Yes	Pass
2	35	34.8	40.2	40.0	37.1	43.3	-0.2	Yes	Yes	Yes	Pass
3	35	34.6	38.8	40.8	35.7	41.8	2.0	Yes	Yes	No	
4	35	34.7	38.8	41.3	35.7	41.8	2.5	Yes	Yes	Yes	Pass
5	35	34.5	38.8	40.8	35.7	41.8	2.0	Yes	Yes	No	
6	35	34.7	38.8	40.6	35.7	41.8	1.8	Yes	Yes	Yes	Pass
7	35	34.7	38.8	40.9	35.7	41.8	2.1	Yes	Yes	No	
8	35	34.6	38.8	41.4	35.7	41.8	2.6	Yes	Yes	No	
9	35	34.6	38.8	40.3	35.7	41.8	1.5	Yes	Yes	Yes	Pass
10	35	34.7	38.8	40.1	35.7	41.8	1.3	Yes	Yes	Yes	Pass
11	35	34.6	38.8	41.4	35.7	41.8	2.6	Yes	Yes	No	
12	35	34.4	38.8	38.8	35.7	41.8	0.0	Yes	Yes	No	

*Includes illumination of intersection icon for a minimum of six consecutive seconds and the reception of sufficient SPaT packets from both intersections (at least 15% of the total).

2.6.2. Multiple Intersection Tests for Nuisance Warning

This test sought to confirm that the presence of the second (alternate) intersection would not result in inappropriate warnings at the first intersection. The test setup (including intersection geometry and nominal speed) for this set of tests was identical to that of the previous set except that the SPaT for the main intersection was broadcasting that all of its approaches had green signals. The approaches for the alternate intersection were all red. Thus, the appropriate system response was for there to be no warning as the test vehicle approached and entered the main intersection. As in the previous set, there were twelve runs. In six of the twelve runs, the number of packets from the alternate intersection was insufficient. In another run, the "Intersection Equipped" icon (which verifies the system is approaching an intersection with an identified geographical intersection description) was not illuminated for six consecutive seconds during the time between the start of the test and the crossing of the stop bar. All seven of these runs were declared invalid. The remaining five valid runs all passed the test. The results are shown in Table 14.

Table 14. Results for Multiple Intersection Nuisance Warning Tests

Run	Nominal Test Speed [mph]	SPaT correct?	Any Warning Observed?	Test Conditions Valid*?	Pass/Fail
1	35	Yes	No	No	
2	35	Yes	No	Yes	Pass
3	35	Yes	No	Yes	Pass
4	35	Yes	No	Yes	Pass
5	35	Yes	No	No	
6	35	Yes	No	No	
7	35	Yes	No	No	
8	35	Yes	No	No	
9	35	Yes	No	Yes	Pass
10	35	Yes	No	No	
11	35	Yes	No	No	
12	35	Yes	No	Yes	Pass

*Includes illumination of intersection icon for a minimum of six consecutive seconds and the reception of sufficient SPaT packets from both intersections (at least 15% of the total).

2.7. Dynamic Signal Change Tests

The CICAS-V system must be able to react appropriately when a signalized intersection changes its signal setting while the vehicle is in or near the optimum warning range. A series of tests was devised to verify that warnings would be issued when appropriate but not when they were not called for.

2.7.1. Dynamic Signal Change to Yellow (Too Late to Warn)

The first set of tests examined the case in which signals for the approach lanes transitioned to yellow too late for an effective warning. The traffic control device on the test track was configured to change from green to yellow as the test vehicle approached the nominal warning distance and remain yellow until after the vehicle passed the stop bar (so long as it continued at the nominal speed). The geometry of the intersection is shown in Figure 7.

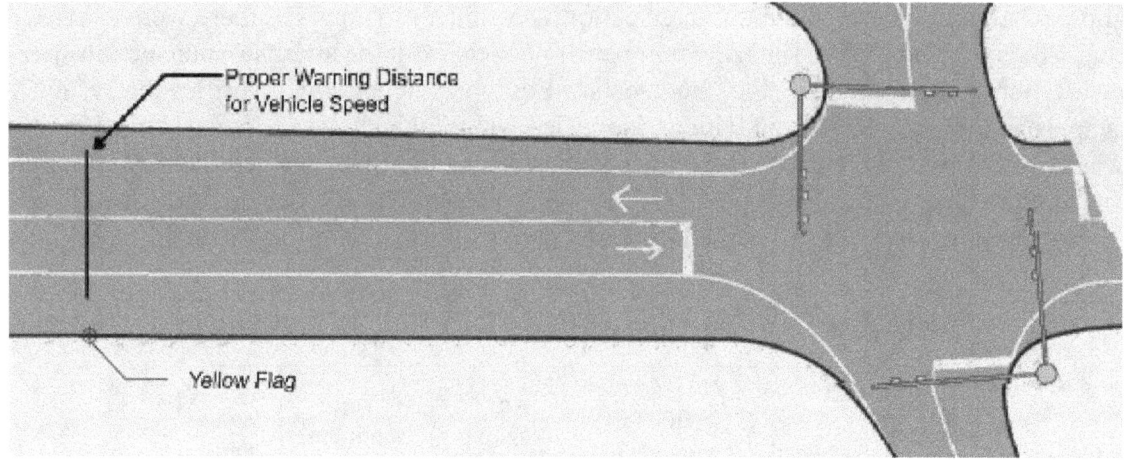

Figure 7. Geometry for Dynamic Signal Change Warning Tests

The software of the traffic control device triggered the change to yellow based on the time-to-intersection (TTI) calculated on the vehicle. The TTI trigger algorithm considered the total latency involved in the test vehicle sending the command and the roadside equipment receiving it and implementing it. There was also a margin of safety to assure the vehicle would visibly clear the stop bar before the transition to red as well as to account for cruise control speed variability and other unknowns. The TTI that triggered the transition to yellow will increase with the latency. If the yellow duration is fixed, the margin would decrease the trigger TTI. The trigger TTI can be calculated as:

Trigger TTI = (yellow duration) + latency - margin

The latency was found experimentally to be approximately 0.2 sec. The safety margin was set at 0.4 sec, the time to travel about one car length at the nominal speed of 35 mph. The duration of the yellow light was set at 3.6 sec, which is a reasonable setting for a 35 mph approach intersection. Thus, the trigger TTI for this set of tests was 3.4 sec. This value satisfied the test requirement that the transition to yellow occurs before the nominal warning location. Even at the highest valid test speed of 37.5 mph (60.4 km/h), the nominal warning distance would have been 46.27 m. A valid warning could occur 0.2 sec (3.34 m) before that point, or 49.61 m from the stop bar. After the latency period, the expected transition would occur 3.2 sec (53.64 m) before entering the intersection. The time sequence of this test is shown schematically in Figure 8.

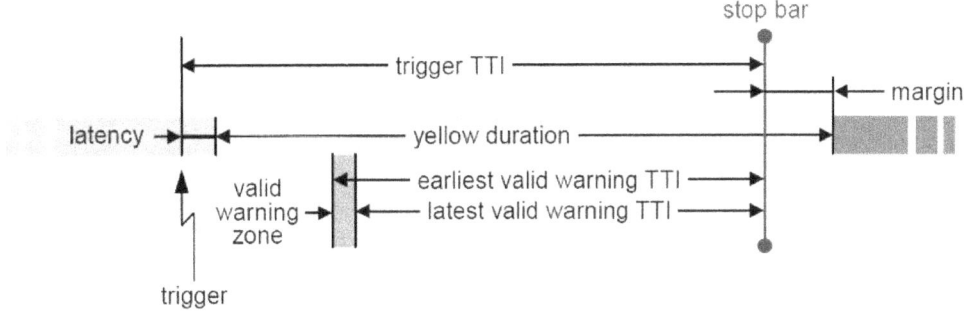

Figure 8. Time Sequence Schematic for Dynamic Signal Change to Yellow Tests

For a test in this set to be considered valid, the DAS must have confirmed that the transition to yellow occurred before the vehicle reached the valid warning zone and the transition to red occurred after the vehicle passed the stop bar. A valid test was considered to have been passed if no warning occurred and to have failed if any warning occurred. All ten tests in this set were valid and passed. The results of this set of tests are shown in Table 15.

Table 15. Results for Dynamic Signal Change to Yellow Tests

Run	Nominal Test Speed [mph]	Change to Yellow Before Warning Zone?	Change to Red After Stop Bar?	Test Conditions Valid?	Any Warning Observed?	Pass / Fail
1	35	Yes	Yes	Yes	No	Pass
2	35	Yes	Yes	Yes	No	Pass
3	35	Yes	Yes	Yes	No	Pass
4	35	Yes	Yes	Yes	No	Pass
5	35	Yes	Yes	Yes	No	Pass
6	35	Yes	Yes	Yes	No	Pass
7	35	Yes	Yes	Yes	No	Pass
8	35	Yes	Yes	Yes	No	Pass
9	35	Yes	Yes	Yes	No	Pass
10	35	Yes	Yes	Yes	No	Pass

2.7.2. Dynamic Signal Change to Red (Sufficient to Warn)

This test sought to confirm that the CICAS-V system would alert the driver if it calculated that the traffic signal will change to red before the test vehicle reaches the stop bar, even if the traffic signal were still yellow when the vehicle reaches the optimum warning distance. The intersection geometry shown in Figure 9 is essentially the same as in the previous set of tests, but the timing of signal changes is different.

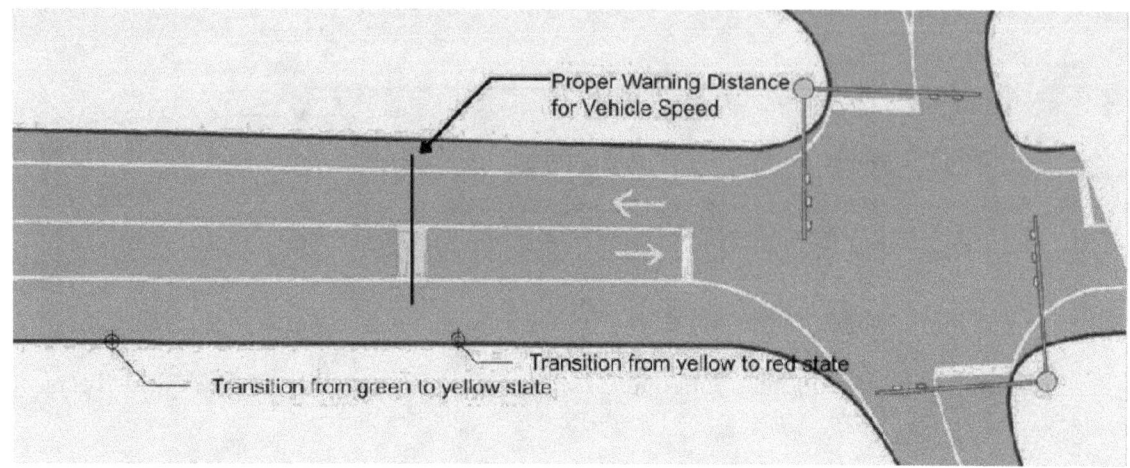

Figure 9. Geometry for Dynamic Signal Change to Red (Sufficient to Warn) Tests

At the beginning of each test run, the intersection traffic signal was green. The signal transitioned to yellow before the optimum warning distance for the nominal speed. The signal transitioned to red before the test vehicle reached the stop bar and remained red at least until the test vehicle reached the stop bar. In order for the system to pass a test run, it had to issue a warning in the appropriate range. Note that in the test protocol the test vehicle did not brake but retained the nominal speed when the traffic signal transitioned to yellow. The time sequence of this test is shown schematically in Figure 10. The nominal test speed is 35 mph.

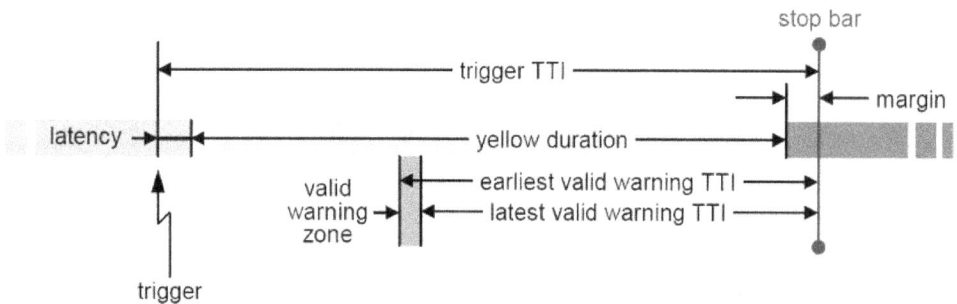

Figure 10. Time Sequence Schematic for Dynamic Signal Change to Red (Sufficient to Warn) Tests

The signal change was triggered based on the test vehicle's calculated TTI. As with the previous set of tests, the TTI trigger allowed for the system latency between sending the command and implementing it. A margin of safety should allow for variations in cruise control speed as well the time to assure the transition to red occurred before the test vehicle crossed the stop bar. The value of TTI that triggers the transition to yellow would thus be:

$$\text{Trigger TTI} = (\text{yellow duration}) + \text{latency} + \text{margin}$$

The magnitudes of these components are the same as in the previous set of tests, but because of the sign change on the margin, the trigger value of TTI was 4.0 sec.

A practical maximum to the trigger value of TTI would be that value that would cause the transition to red while the test vehicle was in the optimum warning range. The most critical value for that would be if the actual test speed were 32.5 mph (52.3 km/h), in which case the warning would occur 2.4 sec before the stop bar.

A test run would have been considered invalid if the traffic signal did not change to red between the time the vehicle passed the optimum warning range and the time it crossed the stop bar. The average warning deviation was 0.0 m with a standard deviation of 0.7 m. A total of ten test runs were conducted. All ten runs were valid and each test passed. Table 16 summarizes the results.

2.7.3. Dynamic Signal Change to Green (No Warning Necessary)

Finally, the CICAS-V system should not produce a warning if the traffic signal will turn to green before the test vehicle arrives at the optimum warning location. In this set of tests, the test vehicle approached the intersection with the traffic signal red. The traffic signal changes to green before the test vehicle reaches the optimum warning distance and remained green at least until the test vehicle passed the stop bar. The geometry for this test is shown in Figure 11. The nominal test speed was 35 mph.

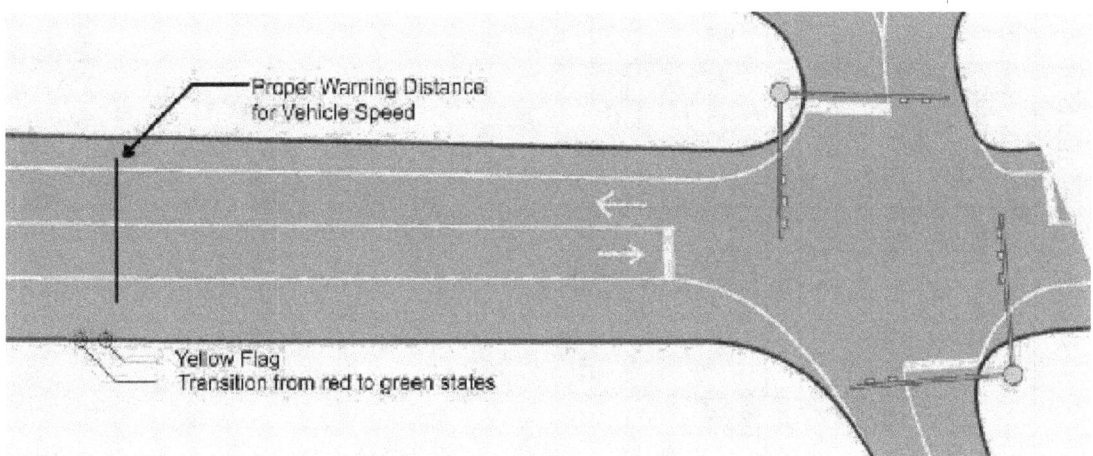

Figure 11. Geometry for Dynamic Signal Change to Green (No Warning Necessary) Tests

Table 16. Results for Dynamic Signal Change to Red (Sufficient to Warn) Tests

Run	Nominal Test Speed [mph]	Recorded Test Speed [mph]	Nominal Warning Distance [m]	Actual Warning Distance [m]	Minimum Warning Distance [m]	Maximum Warning Distance [m]	Warning Distance Deviation [m]	Transition to Red in Correct Range?	All Warnings Observed?	Test Conditions Valid?	Pass / Fail
1	35	34.4	38.8	37.9	35.7	41.8	-0.9	Yes	Yes	Yes	Pass
2	35	34.4	38.8	37.8	35.7	41.8	-1.0	Yes	Yes	Yes	Pass
3	35	34.5	38.8	39.2	35.7	41.8	0.4	Yes	Yes	Yes	Pass
4	35	34.4	38.8	39.4	35.7	41.8	0.6	Yes	Yes	Yes	Pass
5	35	34.4	38.8	39.7	35.7	41.8	0.9	Yes	Yes	Yes	Pass
6	35	34.5	38.8	38.9	35.7	41.8	0.1	Yes	Yes	Yes	Pass
7	35	34.3	38.8	38.3	35.7	41.8	-0.5	Yes	Yes	Yes	Pass
8	35	34.6	38.8	39.2	35.7	41.8	0.4	Yes	Yes	Yes	Pass
9	35	34.5	38.8	38.6	35.7	41.8	-0.2	Yes	Yes	Yes	Pass
10	35	34.5	38.8	39.3	35.7	41.8	0.5	Yes	Yes	Yes	Pass

Once again, the phase change was triggered when a critical value of TTI (as calculated on the test vehicle) was reached. The maximum TTI of a warning in a valid test would be 2.78 sec for a 37.5 mph approach. As previously stated, the latency associated with the transmission, reception, and implementation of the command to change the signal was taken to be 0.2 sec. Similarly, the latency for the on-board equipment to receive, interpret, and react to the traffic signal phase change was also taken to be 0.2 sec. Hence, the value of TTI that triggered the phase change from red to green was set to 3.18 sec (2.78 sec + 0.2 sec + 0.2 sec). The time sequence of this test is shown schematically in Figure 12. A test will be considered valid if the traffic signal transitions from red to green before the earliest valid warning distance.

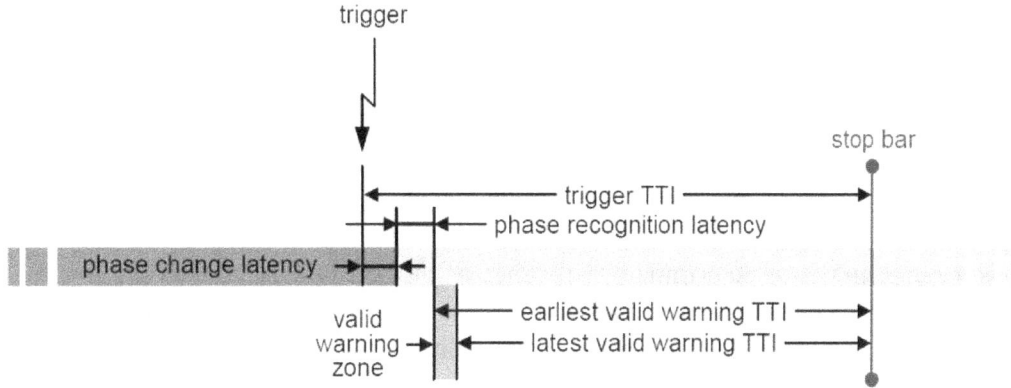

Figure 12. Time Sequence Schematic for Dynamic Signal Change to Green (No Warning Necessary) Tests

The system was considered to have passed for a given valid run if it did not issue any warning; any warning would have constituted failure of the run. Eight tests runs were conducted for this set of tests. All eight runs were valid. The system passed all eight tests. The details are summarized in Table 17.

Table 17. Results for Dynamic Signal Change to Green (No Warning Necessary) Tests

Run	Nominal Test Speed [mph]	Change to Green Before Warning Zone?	Test Conditions Valid?	Any Warning Observed?	Pass / Fail
1	35	Yes	Yes	No	Pass
2	35	Yes	Yes	No	Pass
3	35	Yes	Yes	No	Pass
4	35	Yes	Yes	No	Pass
5	35	Yes	Yes	No	Pass
6	35	Yes	Yes	No	Pass
7	35	Yes	Yes	No	Pass
8	35	Yes	Yes	No	Pass

2.8. Signal Phase and Timing Reception and Reflection Tests

The SPaT and DSRC functions of the CICAS-V system were evaluated for their ability to function under adverse conditions. The tests of these capabilities were not objective tests of the system's ability to whether or not to issue an appropriate warning, but engineering tests of the hardware in the context of assessing the robustness of the overall system. As an engineering test, there were no pass/fail criteria.

The functionality of the system depends on the test vehicle's ability to receive and interpret timely messages from the SPaT (as well as DGPS and GID) through the DSRC. If communications were disrupted, appropriate signals may not be received and evaluated. To test this capability, the test vehicle was driven closely behind a tractor-trailer while approaching an intersection in which a warning was expected. The CICAS-V system was expected to produce an appropriate warning despite the presence of the truck so long as the minimum requirements for packet reception were met.

In this set of tests, the test observer used an onboard equipment interface to clear the GID from the system and institute the standard initialized state. The observer used a hand-held laser range finder to determine the distance of the test vehicle behind the tractor-trailer. The SPaT server transmitted a red traffic signal for all approaches to the intersection. The tractor-trailer approached the intersection at 35 ± 5 mph. With input from the laser range finder via test observer, the driver of the test vehicle maintained a distance of 3 to 6 m behind the tractor-trailer as shown in Figure 13. The maintenance of a distance of less than 6 m (as measured by the DAS radar) was required for a valid test.

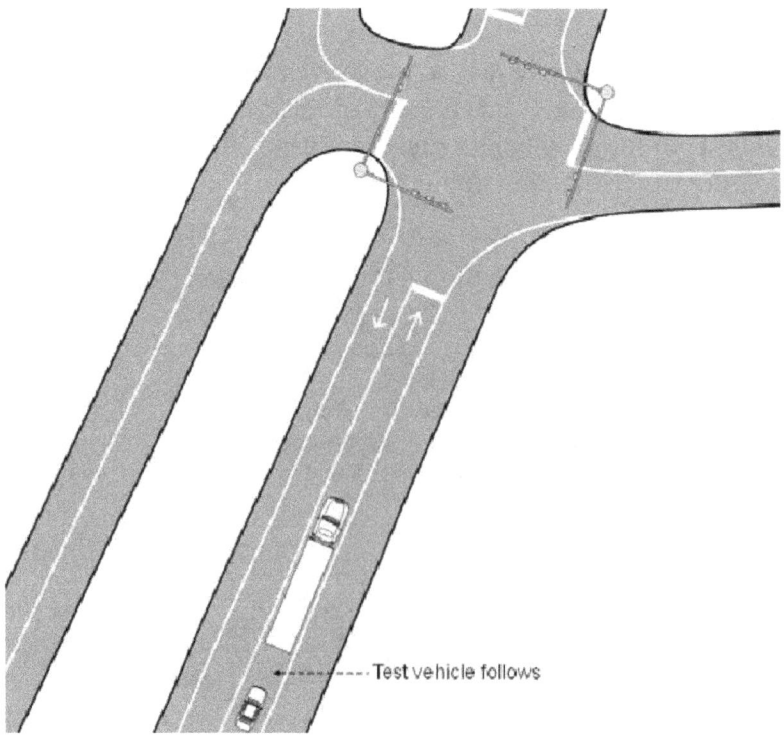

Figure 13. Geometry for SPaT Reflection and Reception Tests

The system was expected to give a normal warning in the expected warning range despite receiving the GID significantly later in each test run. Eight runs were performed. All eight runs were valid and recorded warnings in the appropriate range. The average warning distance deviation was +1.5 m with a standard deviation of 0.8 m. Table 18 provides detailed results of this set of tests.

2.9. Objective Test Conclusions

The results of these series of tests conclusively show that the CICAS-V system behaves as designed in a wide variety of common driving situations. The test vehicle consistently warned the driver when the vehicle was exceeding the target speed for a safe stop in a lane designated to stop, whether by a stop sign or by a traffic signal, over a large range of test speeds. The system consistently distinguished between the required alarm state for the current lane and that of nearby lanes and was sufficiently robust even if the vehicle were located at the edge of the designated lane or dynamically shifted between lanes in which the appropriate alert status changed. It also differentiated between multiple intersections in close proximity and engaged a warning state appropriate for the relevant intersection and lane. Finally, the engineering test demonstrated significant robustness in the ability of the system to evaluate the situation and warn correctly under conditions that severely inhibited the line-of-sight wireless reception. Through this series of tests, the U.S. Department of Transportation, VTTI, and CAMP have demonstrated diligence in verifying that the CICAS-V system can perform in a consistent and repeatable manner.

Table 18. Results of SPaT Reflection and Reception Tests

Run	Nominal Test Speed [mph]	Recorded Test Speed [mph]	Nominal Warning Distance [m]	Actual Warning Distance [m]	Minimum Warning Distance [m]	Maximum Warning Distance [m]	Warning Distance Deviation [m]	Radar Distance < 6m?	All Warnings Observed?	Test Conditions Valid?
1	35	35.2	40.2	41.5	37.1	43.3	1.3	Yes	Yes	Yes
2	35	35.7	41.7	42.0	38.5	44.9	0.3	Yes	Yes	Yes
3	35	35.6	41.7	43.3	38.5	44.9	1.6	Yes	Yes	Yes
4	35	35.2	40.2	42.7	37.1	43.4	2.5	Yes	Yes	Yes
5	35	34.4	38.8	41.3	35.7	41.8	2.5	Yes	Yes	Yes
6	35	35.9	41.7	42.4	38.5	44.9	0.7	Yes	Yes	Yes
7	35	35.9	41.7	42.9	38.5	44.9	1.2	Yes	Yes	Yes
8	35	35.8	41.7	43.8	38.5	44.9	2.1	Yes	Yes	Yes

3. PILOT TEST

3.1. Overview

In a second series of evaluations, the CICAS-V system was employed in a pilot test in which volunteer drivers evaluated the system on public roads and, in some cases, at a test track facility [1]. CAMP provided two identically-equipped 2006 Cadillac STS vehicles that were equipped with several advanced safety devices, including the CICAS-V. The tests were conducted by VTTI in the Blacksburg/Christiansburg, VA area. A total of 93 test subjects were recruited. VTTI reported occasional failures of the hardware that translated controller-area network bus signals into the appropriate format. This resulted in the loss of about 5 percent of the data, resulting in usable data from only 87 test subjects. The 87 subjects were roughly evenly distributed by gender and across three age categories (18-30, 35-50, 55+), as shown in Table 19.

Table 19. Distribution of Evaluated Drivers by Age and Gender

Age Group	Gender		Total
	Male	Female	
18-30	17	15	32
35-50	10	14	24
55+	15	16	31
Total	42	45	87

3.2. Pseudo-Naturalistic Test Results

The volunteer drivers were briefed on the advanced safety features available in the test vehicles. The CICAS-V system was not emphasized more than any of the other systems. Using a navigation system to guide them, the participants drove the vehicle unaccompanied through along a pre-planned 36-mile course that traversed three signalized and ten stop sign-controlled intersections equipped with CICAS-V roadside equipment. As the route traversed several of these intersections multiple times, it included a total of 20 maneuvers at signalized intersections and 32 maneuvers at stop sign-controlled intersections. The distribution is given in Table 20. The course took about two hours to complete. Figure 14 shows a map with some of the equipped intersections.

Table 20. Turn Maneuvers by Intersection Type in Pseudo-Naturalistic Study

Signalized Intersections [Three Locations]				
Permissive Left	Protected Left	Straight	Right	Total
2	5	11	2	20
Stop Sign-Controlled Intersections [Ten Locations]				
Left		Straight	Right	Total
12		6	14	32

Image is the copyrighted work of Microsoft® and subject to the terms and conditions of the Microsoft® license agreement.

Figure 14. Map of Pseudo-Naturalistic Study Route with Labeled Intersections

During the first several days of the pseudo-naturalistic trials, it was noticed that fourteen of fifteen drivers had received alerts clustered at five stop sign-controlled intersections (Table 21). The approaches to all five of these intersections had a 3.8 to 7 percent uphill grade. The algorithm used for warning at such intersections, designated "Stop-Controlled Algorithm 1", considered the status of the brakes in determining whether to issue an alert. If a driver was pressing the brake, the design of the warning logic assumed the driver was attentive to the intersection and the alert was suppressed. On uphill grades, drivers tended to press the brake later in their approach, using gravity to slow the vehicle. Since the algorithms were developed on flat intersection approaches, the later braking caused the warning to activate more often than was expected.

Table 21. Distribution of "Stop-Controlled Algorithm 1" Drivers by Age and Gender

Age Group	Gender		Total
	Male	Female	
18-30	2	1	3
35-50	1	4	5
55+	4	3	7
Total	7	8	15

For the remaining 72 drivers, the stop sign-controlled algorithm was replaced with "Stop-Controlled Algorithm 2". Instead of using the brake pedal status, the Stop-Controlled Algorithm 2 monitored the deceleration level of the vehicle to determine whether to suppress the alert. After the implementation of the new algorithm, the percentage of drivers receiving stop sign-controlled alerts dropped from 93 percent (14 of 15) to four percent (3 of 72).

The three valid stop sign-controlled alerts received from the Stop-Controlled Algorithm 2 were all experienced at the same intersection. The course called for a straight crossing of an intersection where the stop sign was partially obscured at a distance. None of the drivers exhibited an indication of stopping before the alert. Peak decelerations ranged from 0.46 g to 0.60 g. Average decelerations ranged from 0.33 g to 0.41 g. This information indicates these were likely valid and necessary alerts that may have prevented these drivers from violating the stop sign.

The traffic signal algorithm was consistent for all 87 participants evaluated. A total of seven alerts were received at traffic signal-controlled intersections. A post-test analysis revealed that six of these seven alerts were in fact invalid. Four of these alerts were the result of an error in the GID of one lane of one intersection. One lane for straight-ahead traffic was coded as a protective left turn lane, resulting in an erroneous alert when the vehicle tried to proceed through the intersection under a green signal while the actual left turn lane was under a red signal. The error was recognized after the first occurrence, but researchers decided to allow the GID to learn about drivers' responses when receiving a false alert at a green traffic signal. The first three drivers did in fact promptly evaluate the situation and ignore the alert, but the fourth stopped abruptly despite the green signal. The researchers immediately remedied the erroneous GID to avoid any unexpected driver reactions.

The two other invalid alerts were related to emergency vehicle signal preemption. The firmware in the traffic signal can give priority green signals to emergency vehicles. In both of these cases, such an event occurred shortly before the approach of the test vehicle. The firmware did not update the CICAS-V roadside equipment as expected, resulting in the RSE broadcasting an incorrect signal status and the subsequent invalid alert.

The valid alert occurred as a middle-aged male driver approached an intersection to execute a straight-crossing maneuver. The driver slowed and stopped after receiving the warning. As the vehicle in front of him was crossing over the stop bar as the signal turned red, a decision to proceed through the intersection would have unambiguously resulted in a violation.

3.3. Test Track Results

Twenty-three subjects participated in the test track study on the VTTI Smart Road. On the test track, an experimenter accompanied the subjects. In the final intersection approach on the test track, each subject was distracted using a specific VTTI protocol in order to trigger a CICAS-V alert [1]. Five of the subjects were not fully distracted by the protocol and thus did not provide the desired data on the system utility. The demographic distribution of the remaining eighteen drivers is given in Table 22.

Table 22. Distribution of Sufficiently Distracted Test Track Drivers by Age and Gender

Age Group	Gender		Total
	Male	Female	
18-30	3	3	6
35-50	2	4	6
55+	3	3	6
Total	8	10	18

The general format of the test track study was to make ten passes through the intersection while adjusting various dashboard controls. The subjects were told that the purpose of the tests was to evaluate the human factors aspects of various control configurations. Another vehicle, ostensibly for facility maintenance, made carefully choreographed crossings of the intersection which demonstrated that the traffic signal would be changed to give the maintenance vehicle a green signal. On the final run, another confederate vehicle followed the test vehicle and the traffic signal was changed from green to yellow just as the driver was tasked to look away from the road to adjust a control setting. If sufficiently distracted, the driver would be given an alert and left to make a split second decision regarding stopping while being closely followed. One driver opted to continue through the intersection while the remaining seventeen made a compliant stop. Thus, the system was effective in preventing intersection violations for 94 percent of drivers who were warned while intentionally distracted and followed closely by another vehicle.

3.4. Pilot Test Questionnaires

After completing the testing, subjects were given questionnaires. The specific questions that were included were dependent on whether or not they received an alert, what kind of alert they received, whether it was received in the pseudo-naturalistic course or on the Smart Road test track. Not all participants filled out the questionnaires. Table 23 describes the five groups by alerts experienced.

The drivers who were alerted by Stop-Controlled Algorithm 1 tended to experience many alerts (an average of nearly four each for the thirteen drivers who completed the questionnaire). They said they found the alerts useful and effective at gaining the driver's attention. They also found the alerts annoying when deemed unnecessary and indicated that they often resulted in braking without checking for following vehicles. The drivers also noted that the visual components of the system (blue icon for intersection detected, red icon for alert) were less noticeable and effective than the audible and haptic components.

Table 23. Distribution of Questionnaire Respondents by Alerts Experienced

Alert Group	Respondents	Potential Respondents
Valid Alert - Stop Controlled Algorithm 1*	13	14
Valid Alert - Stop Controlled Algorithm 2	3	3
Invalid Alert - Traffic Signal Violation**	6	6
Valid Alert - Smart Road Test Track Only	17	17
No Alert Experienced	47	49
Total	86	89

*One subject with Stop Controlled Algorithm 1 experienced no alert
**Included one subject with a valid traffic signal alert and one subject who participated in the Smart Road protocol. None received a stop controlled alert.

The three drivers who experienced single valid warnings from Stop-Controlled Algorithm 2 were more favorable in their responses. These subjects did not receive nuisance alarms and were all alerted to the same partially obscured stop sign. Hence, their evaluation of the system tended to be more favorable, as it was perceived as achieving its goal of only warning when the driver was in danger of inadvertently violating an intersection.

The six drivers who received invalid traffic signal alerts (one of whom received a valid traffic signal alert) did not find the system to be overly annoying or distracting. They also found the visual component to be less useful than the audible and haptic components.

None of the seventeen subjects who participated in the Smart Road test and completed the questionnaire received a stop-controlled alert. Thus, all the alert-specific questions were related to the traffic signal-controlled alert experienced on the test track. As all these subjects experienced a valid alert that occurred under foreseeable (if somewhat "surprising") conditions, they all had favorable opinions of the system's utility. They also felt that audible and haptic components were more effective than the visual ones.

The drivers who experienced no alerts were of course not expected to answer questions related to alerts. They were asked primarily questions about the blue icon indicating that they were approaching an instrumented intersection. Some participants did not answer these questions, presumably because they did not notice the icon at all. Most comments indicated that the icon was not very conspicuous. Care should be taken to evaluate the ultimate desirability of a conspicuous visual icon that merely alerts to the presence of an intersection.

3.5. Pilot Test Conclusions

The pilot test program of the CICAS-V system demonstrated that it functions well as implemented on public roads. The system reacted appropriately in the vast majority of the 2,618 stop controlled intersection crossings and the 1,455 signal controlled intersection crossings recorded by the DAS.

There were some nuisance warnings related to Stop-Controlled Algorithm 1, which were immediately remedied by its replacement. The invalid signal controlled warnings caused by an erroneous GID were also shown to be capable of expedient correction. The issue related to synchronization of signal phase after the use of emergency vehicle priority at some intersections was discovered somewhat by serendipity and demonstrates the value of a more-than-cursory pilot test.

Significantly, the CICAS-V system did appropriately warn three drivers who may have inadvertently violated an intersection controlled by a partially obscured stop sign and one driver who might have otherwise violated a red traffic signal. It also appropriately warned all eighteen (intentionally) distracted drivers on the Smart Road test track, facilitating seventeen of them to avoid a violation. As the system evolved, its ability to issue effective warnings and minimize unintentional driver irritation has improved.

4. OVERALL RECOMMENDATIONS FOR CICAS-V SYSTEM

The results of the twelve sets of objective tests and the two phases of the pilot test indicate that the CICAS-V system is ready for deployment in a field operational test. It was able to demonstrate in the objective tests that it could resolve potentially challenging geometric and dynamic situations and still issue appropriate warnings and avoid nuisance alerts. The pseudo-naturalistic tests and Smart Road tests portions of the pilot test clearly demonstrated the ability of the system to appropriately alert naïve drivers of the impending risk of violation of a traffic control device.

The program also demonstrates the practical need for the fine-tuning of a system before commencing a full-scale field test, particularly in the actual region where the field test will be conducted. Issues with erroneous or ambiguous GIDs (including GIDs in close proximity) should be discovered and remedied as expediently as possible. Other more basic system functions (the ability to account for emergency vehicle priority mechanisms, the conspicuity or elimination of visual alerts) may be revised before a field test and would require due diligence in verifying intended functionality. It is also possible that other unforeseen issues (akin to the effect of local topography on Stop-Controlled Algorithm 1) will need to be addressed. Thus, it is recommended that a pilot and test track program be undertaken again before releasing the revised system into a full-scale field trial.

Finally, the reliability of data acquisition and recording hardware should be resolved. The requirements for a full field operational test will be more stringent than for the pilot testing and should be verified in advance.

5. REFERENCES

[1] Neale, V. L., Doerzaph, Z.R., Viita, D., Bowman, J., Terry, T., Bhagavathula, R., and Maile, M. *Cooperative Intersection Collision Avoidance Systems Limited to Stop Sign and Traffic Signal Violations (CICAS-V) Subtask 3.4 Interim Report: Human Factors Pilot Test of the CICAS-V.* Appendix A: Task 3 Final Report – Human Factors Development and Testing. *Cooperative Intersection Collision Avoidance System Limited to Stop Sign and Traffic Signal Violations (CICAS-V) – Phase I Final Report.* Washington, DC: Federal Highway Administration, FHWA-JPO-10-068, September 2008.

[2] Maile, M., Ahmed-Zaid, F., Basnyake, C., Caminiti, L., Kass, S., Losh, M., Lundberg, J., Masselink, D., McGlohon, E., Mudalige, P., Pall C., Peredo, M., Popovic, Z, Stinnett, J., and VanSickle, S. *Cooperative Intersection Collision Avoidance System Limited to Stop Sign and Traffic Signal Violations (CICAS-V) Task 7 Final Report: Objective Test Procedures.* Appendix H-1: Task 7 – Final System Test Plan and Procedures. *Cooperative Intersection Collision Avoidance System Limited to Stop Sign and Traffic Signal Violations (CICAS-V) – Phase I Final Report.* Washington, DC: Federal Highway Administration, FHWA-JPO-10-068, September 2008.

[3] Perez, M. A., Neale, V. L., Kiefer, R. J., Viita, D., Wiegand, K., and Maile, M. *Cooperative Intersection Collision Avoidance Systems Limited to Stop Sign and Traffic Signal Violations (CICAS-V) Subtask 3.3 Interim Report: Test of Alternative Driver-Vehicle Interfaces on the Smart Road.* Appendix A: Task 3 Final Report – Human Factors Development and Testing. *Cooperative Intersection Collision Avoidance System Limited to Stop Sign and Traffic Signal Violations (CICAS-V) – Phase I Final Report.* Washington, DC: Federal Highway Administration, FHWA-JPO-10-068, September 2008.

[4] Maile, M., Ahmed-Zaid, F., Basnyake, C., Caminiti, L., Kass, S., Losh, M., Lundberg, J., Masselink, D., McGlohon, E., Mudalige, P., Pall C., Peredo, M., Popovic, Z, Stinnett, J., and VanSickle, S. *Cooperative Intersection Collision Avoidance System Limited to Stop Sign and Traffic Signal Violations (CICAS-V) Task 11 Final Report: Objective Tests.* Appendix H-2: Task 11– Final Report Objective Tests. *Cooperative Intersection Collision Avoidance System Limited to Stop Sign and Traffic Signal Violations (CICAS-V) – Phase I Final Report.* Washington, DC: Federal Highway Administration, FHWA-JPO-10-068, September 2008

DOT HS 811 499
July 2011

www.ingramcontent.com/pod-product-compliance
Lightning Source LLC
Chambersburg PA
CBHW081759170526
45167CB00008B/3248